John Tarilanyo Afa

# Substation Faults and Protection

GRIN Verlag

**Bibliografische Information der Deutschen Nationalbibliothek:**

Die Deutsche Bibliothek verzeichnet diese Publikation in der Deutschen Nationalbibliografie; detaillierte bibliografische Daten sind im Internet über http://dnb.d-nb.de/ abrufbar.



**Imprint:**


Druck und Bindung: Books on Demand GmbH, Norderstedt Germany
ISBN: 978-3-656-41233-5

**This book at GRIN:**

http://www.grin.com/en/e-book/213055/substation-faults-and-protection

**JOHN TARILANYO AFA**

# CURRICULUM DESIGN

# TECHNICAL REPORT

# SFP 673:

# SUBSTATION FAULTS AND PROTECTION

**ATLANTIC INTERNATIONAL UNIVERSITY**

**HONOLULU, HAWAII**

**2011**

Atlantic International University

## TABLE OF CONTENT

Atlantic International University

**SUBSTATION FAULTS AND PROTECTION**

## 1.0 INTRODUCTION:

The purpose of this work is to establish the type of faults that may result when substations are sited in the region of the Niger Delta due to the prevailing climatic conditions. This will aimed at ensuring as far as possible a small probability of damage to substation insulations.

For most transmission lines, relatively large numbers of yearly flashovers are permitted but such number of insulation damages is absolutely not allowed for substations. Flashover of insulation at substation means a short circuit on the busbar which even with the modern means of relay protection can cause most sever system damages. Substations unlike lines have very low probability of damage therefore a quantitative idea of the probability, the so called index of lightning resistance of a substation is used. It is equal to the calculated number of years during which a voltage dangerous for the substation installation does not occur. For modern high voltage substations, the index of lightning resistance is calculated as hundred or even as thousand years which is a proof of attempts made by designers to ensure the largest degree of lightning resistance of substation (Rao,2008,U.S Dept of Agri.2001,Martinez and Castro 2003).

Substations must be protected from the direct lightning strokes and voltage waves travelling from the line as well as switching surges. In transmission lines the induced strokes (indirect strokes) due to lightning are important for 11kv lines only. For high voltage transmission lines (up to 220kv) the surges due direct lightning strokes determine the line insulations design. For extra high voltage (400kv and above) the severity of swiching surges is much more than that due to lightning (Gupta, 2008). Whether external or internal over voltage, for reliable operations of substation proper protection is necessary. There are factors that may not produce over voltage but are capable of affecting the operation of the equipments as well as lowering the flashover, these are humidity, rain temperature, pressure and to some extent contamination on the surface of insulator and equipments (outdoor).

For proper consideration of these factors with the external and internal over voltages, a proper selection of protective devices may be made for the safe operation of substation

## 2.0 GENERAL ANALYSIS AND BACKGROUND THEORY:

Substations can simply be seen as a combination of apparatus that transforms the characteristics of electrical energy from one form or level to another form with the provision of facilities for switching. There are various types of substations and are classified according to their services, design, voltage level and functions. Whatever function a substation is meant for, it is the most sensitive part of the supply system.

For any reliable operation of substation the basic functions must be performed.

- Protection against direct stroke
- Protection over travelling surges
- Proper earthing scheme must be maintained

**2.1 Protection Against Direct Over Voltage:** The effectiveness of lightning protection schemes for high and extra high voltage (ehv) stations depends upon the degree of overhead shielding against direct strokes to the station area. Quantitatively, information on shielding efficiency in terms of shielding failure exposures as a function of shielding angle has been given in several literatures.

The two ways of protecting the equipment from direct stroke is one of the following ways:

(i) Overhead shielding screen (earthed) covering the outdoor substation and the overhead lines approaching the substation.

(ii) Lightning masts installed at strategic location in the station. Lightning masts are preferred for outdoor stations up to 33kv. For 66kv and above the lightning masts become too tall and uneconomical. The overhead shielding wires are preferred because they give adequate protection and the height of structures in the station provided with overhead shielding wires is comparatively less than that of lightning masts (Rao, 2008 Mousa 1991).

The height above the surface of ground at which the leader discharge finally orients itself on one of the objects on earth is called "the height of orientation of lightning H", which in the first instance depend on the height of the lightning conductor h. It is customary to consider that for lightning conductor up to a height of 30m will be, H = kh, where the proportionality k has a value of 10-20 (Razevig 2003 Uppal and Rao 2009)

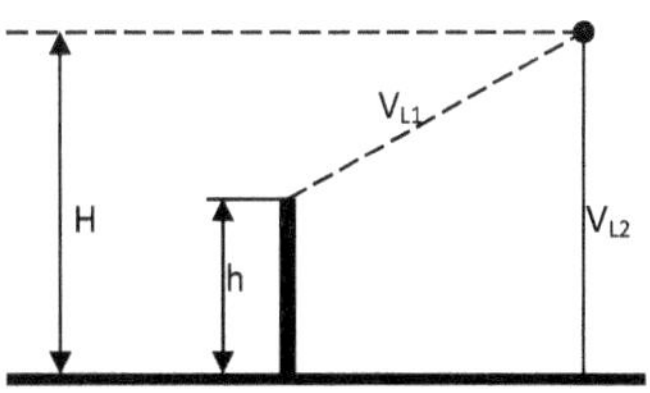


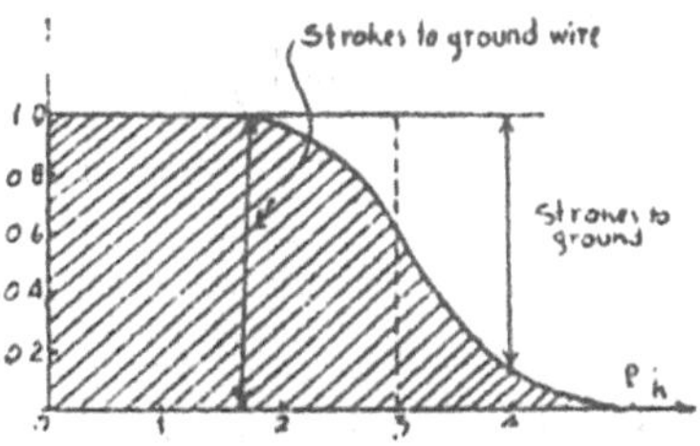


**Fig. 1: Protection zone determination**

Fig.1a shows the point of orientation of lightning that moves along a horizontal line which lies in the same plane as the lightning conductor. When this point is situated directly above the lightning conductor, the discharge will take place on the conductor. However, in proportion to the displacement of the point of orientation from the lightning conductor, the probability of the lightning stroke to the ground increases. The distance will be critical when the breakdown voltages to the lightning conductor $V_{L1}$ and to ground $V_{L2}$ are the same (Razevig,2003, Begamudre, 2008). Fig. 1b shows the probability in terms of distance.

There is therefore an area of protection by shield wire which could be defined in terms of radius or the angle of protection, $\propto$ *and* $\beta$ as shown in fig.2

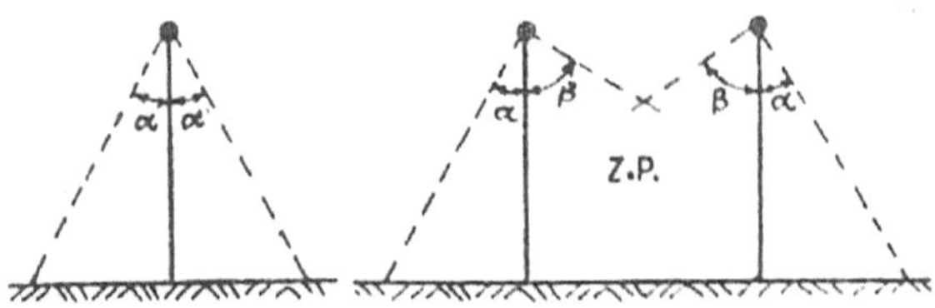


$\propto = 30$ and $\beta = 60$

**Fig. 2: Zone of Protection**

The height of ground wire determines the angle of protection in some cases. Below is given a table of height of ground wire above the girders.

Table 1

| Rated voltage<br><br>KV rms | Minimum ph. to Earth Clearance<br><br>P mm | Minimum Ph to Ph clearance<br><br>E mm | App. Min. Bay Width Wmm<br><br>3 (P + E) | App. Height of Earth Wire above Girder $\propto = 30$<br><br>H mm = W/4 | $\propto 45^{0}$<br><br>H mm = W/2 |
|---|---|---|---|---|---|
| 6.6 | 140 | 180 | 960 | 240 | 480 |
| 11 | 180 | 240 | 1260 | 315 | 630 |

| 33 | 380 | 430 | 2430 | 606 | 1215 |
|---|---|---|---|---|---|
| 66 | 690 | 790 | 4440 | 1110 | 2220 |
| 132 | 1280 | 1480 | 8280 | 2070 | 4140 |
| 220 | 2090 | 2370 | 13380 | 3345 | 6690 |
| 400 | 3070 | 5760 | 26490 | 6622 | 3245 |

*Source: From Elect. Substation Engineering and practice (Rao) pp. 213*

Several techniques have been used including the electro-geometric model technique. With latest knowledge on shielding effect of overhead lightning conductors more accurate estimation of shielding wire failure can be achieved (Uppal and Rao , 2009, Wikipedia 2009)

The following parameters form the framework for the new technique of shielding design.

**1. The Ground Flash Density:**

The frequency of strokes to the station is determined by the number of lightning discharges leaving the cloud area directly above and descending vertically towards the station. In addition some strokes originating in a fringe area near but outside the cloud area may also be attracted to the station once they have progressed far enough downwards. The annual number of strokes to a station is directly proportional to the ground flash density.

**2. Current Amplitude of Stroke to Ground:**

In the electro-geometric model, the termination point of a downward stroke approaching the vicinity of objects on the ground is influenced by the charge in the leader channel and hence by the prospective current amplitude of the return stroke(Ekonomou et al, 2003) Therefore, the chance of a shielding failure also depends upon stroke current amplitude. The probability of the certain amplitude to occur can be expressed in the form of a statistical distribution.

3. **Striking Distance to Elevated Object:**

The striking distance at which the leader is attracted to an object placed on the ground, is related to the change in the leader channel and therefore, also is a function of the prospective current amplitude.

4. **The Leader Approach Angle:**

Observations have shown that many lightning stroke approaching the ground do not follow a straight downward path. This characteristic was presented in the original electro-geometric model by assuming a three dimensional leader angle distribution.

5. **Stroke Termination Point:**

When the lightning stroke leader discharge approaches a station equipped with overhead lightning conductors for direct stroke shielding, it will terminate on that part of the station structure first coming within the striking distance whether the stroke is intercepted by the lightning conductor or it continues until it terminates on another object in the station, depends on the position of the lightning conductor relative to that of the station bus, transformers and other equipment.

The difference in height between the lightning conductors and equipment, the so called effective height of the former and also the horizontal distance between them determines the risk of a shielding failure. Conventionally, this distance and the effective height define the shielding angle. The protected distance and hence the corresponding shielding angle is a function of stroke current amplitude and its statistical distribution(Ekonomou et al, 2003, Uppal and Rao,2009)

6. **The Shielding Failure Risk:**

The chance of a stroke causing a shielding failure may be stated in different ways. The so called exposure is defined as that percentage of strokes to the station not intercepted by the shielding system. If shielding design is based on a constant value of exposure say 1.0% then the actual number of shielding failure is proportional to the station area. In this method, a constant shielding angle corresponding to the exposure is specified, and is selected regardless of the station size.

A more practical and meaningful measure of shielding effectiveness is the shielding failure risk, defined as the number of years in which one shielding failure event is predicted. If a constant value for instance 100 years is adopted for all stations, the designer has to be aware of the fact that in order to meet this requirement, large stations need more effective shielding than small ones. In other words the shielding angle must be reduced as the station area increases (assuming that other factors does not change significantly).

Shielding performance must be related to the overall reliability of the power system, taking into account the economic balance between cost of the lightning protection scheme and the probable loss resulting from shielding failure events.

Once a shielding failure risk has been specified, a tentative decision will be made concerning the shielding method and its overall dimension most suitable for a particular station. The factors that influence such decision are as follows:

- ❖ Choice of vertical or horizontal lightning conductors, electrical, mechanical and aesthetic requirements.
- ❖ Lightning conductor height above ground and effective height above equipment

- Number and preferred location of lightning conductors including supports. These factors are inter-related, therefore a new station design may require several solutions before the optimum shielding system is determined.

### 2.2 Protection Against Travelling Over Voltage

Lightning (surge) arrester are devices that help prevent damage to apparatus due to high voltages. The arrester provides a low impedance path to ground for the current from lightning stroke or transient voltages and then restores to normal operating conditions. Lightning arresters (surge absorbers) are used primarily to protect major equipments like transformers, rotating machines, shunt reactors and even the entire substations (CIGRE 1991,Gustavo et al,2003). Lightning arresters are connected between the line and earth at the substation and power stations. From fig. 3, when the travelling surges reach the diverter and attains the prefixed voltage, a spark is formed across the gap as shown at point P. The diverter then provides a low impedance path to earth through the arrester material. This will hold an IR voltage which depends on the non-linear resistance characteristics of $S_1C$ material (Nagrath and Kothari 2001). These materials by what they are made have a I – V characteristics.

I = $KV^{\alpha}$

I = discharge current

V = voltage across element

$\alpha$ = an exponent more than unity

K = constant

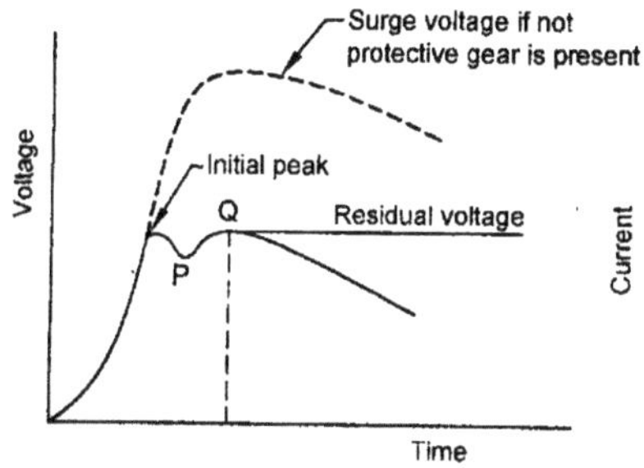


**Fig. 3: V – I Characteristic and Residual voltage IR**

Generally, arresters could be grouped into three categories:

(i) The Rod Gap

(ii) The Expulsion Type

(iii) The Non-Linear Arresters

The Rod Gap Type is simple and less expensive but has the disadvantage of causing dead short circuit. The voltage of phase with respect to ground falls to a very low value. The rod gaps do not exhibits the residual voltage characteristics and the fault continues. Hence rod gaps are no more preferred in modern substations.

The arrester mostly found in the modern use is the non-linear arrester type. They can be considered in three categories:

(i) The Gap type arrester without current limiting functions

(ii) The Gap type arrester with current limiting capability

(iii) The Gapless metal Oxide varistor (MOV)

In fig. 4 the diagram shows the Gap ($S_1C$) type and the metal oxide arrester ($Z_nO$).

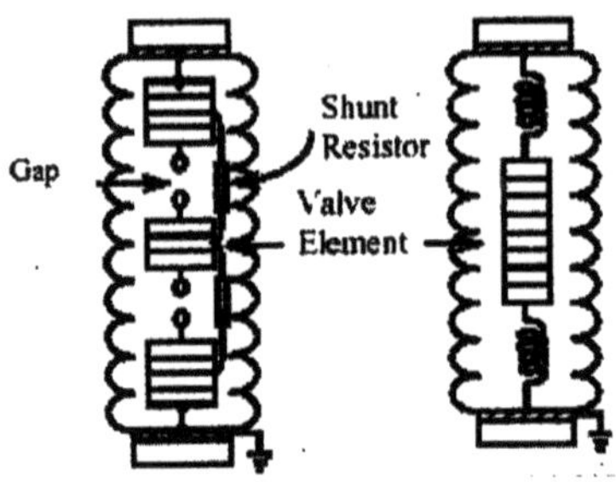


**Fig. 4: $S_1C$ Arrester and $Z_nO$ Arrester**

The various characteristics (I – V) are shown in fig. 5

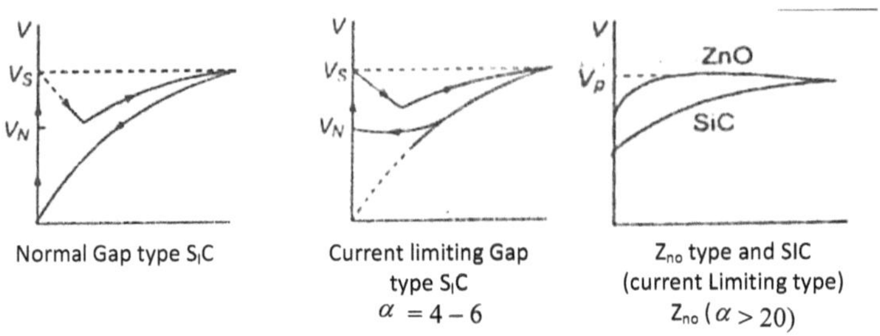


**Fig. 5: Arrester characteristics $Z_nO$ and $S_1C$ types**

## 3.0 METHOD OF STUDY:

Data for climatic and weather conditions were collected from the meteorological station of Federal Minister of aviation. Others were from some environmental study group like Community and Research Development Centre (CREDA) and Lands and people of Rivers State. Personal observations were made from 2010 to date.

From the available records the necessary data were collected and tabulated as shown in table 2 and table 3.

## 4.0 RESULTS

From table 2 the lightning days per year for 2009 and 2010 are shown. About one fifth of lightning are cloud to ground, therefore the lightning days are kept at 80 to 100 days per year. Some weather forecasters kept some coastal areas from 120 to 150 days per year.

**Table 2 Lightning Days for 2009 and 2010**

| Jan. | Feb. | March | April | May | June | July | Aug. | Sept. | Oct. | Nov. | Dec. | Months |
|---|---|---|---|---|---|---|---|---|---|---|---|---|
| 5. | 20 | 43 | 49 | 68 | 35 | 31 | 6 | 40 | 54 | 18 | 7 | Lightning 2010 |
| 7 | 24 | 45 | 71 | 72 | 40 | 29 | 10 | 42 | 61 | 15 | 6 | Lightning 2009 |

The average rainfall, humidity and temperatures are given in table 3.

**Table 3**

| | **Seasonal Rain** | | **Temperature** | | **Relative Humidity** | |
|---|---|---|---|---|---|---|
| Mean Annual mm | Dry Nov. – Dec. mm | Wet March – Oct. mm | Dry °C | Wet °C | Dry mm $H_g$ | Wet mm $H_g$ |
| 2355-3816.8 | 187-370 | 2097-3456.4 | 32-35$^0$C | 24$^0$-28$^0$C | 70-90% | 65-75% |

## 5.0 DISCUSSION:

From the table (table 3) the rainfall is considerably high, in some coastal towns rain falls between 305 – 330 days of the year. The humidity is also high especially from April to September, about 68% to 90% in some areas. Some researchers have placed the lightning day (thunderstorm A) at about 80 to 100 days per year. That is, the ground flash density is high. This means, proper shielding design is needed.

The incoming overhead lines must be properly shielded for at least 2km from the substation to reduce direct stroke on lines (travelling waves).

In the environment considered there is high humidity (68 to 90%) therefore, there may be a likely breakdown of gap below the normal predetermined voltage in some rainy periods and the characteristic of the rod gap which allows follow-up current (dead short circuit) may result to fault situation at some operation time.

Darvenzia, recommended (Dahiya and Attu 2009) that all silicon carbide arresters that have been in service for over 13 years be replaced due to moisture ingress. He test reveals that degradation was evident in 75% of arrester. The situation may even be worst in an environment having at least 200 rainy days in 365 days.

From the performance of the gapless arrester ($Z_nO$), the design eliminates the high heat associated with the arcing discharges as may be the case of the gap arresters ($S_1C$): The MOV arrester has two voltage rating (duty cycle and maximum continuous operation voltage) unlike the silicon carbide that has only the duty cycle rating;. A metal oxide surge arrester utilizing zinc-oxide blocks provides the best performance, as surge voltage level, thereby improving system protection. Failure is reduced, as there is no air gap contamination possibility. Also the leakage current at operating frequency is small (between 1mA to 5mA).

## 6.0 RECOMMENDATIONS:

- Due to the high keraunic level in the area the shielding conductor design must be critical, that is, the protective angle must be made as small as possible and aesthetically allowed (the angle depends on the working area). The system needed masts or shielding wire depends on the designer. But for more than 33kv, the shielding conductor method is recommended (either fixed angle method or rolling sphere method of design).
- The lightning arrester should be installed as close as possible to the transformer to avoid multiple reflections.
- Use of Rod-gap (spark gap) should be minimized due to follow-on current at normal frequency or dead short circuit during its operation. It could only be used as backup protection.
- All climatic factors that may affect the proper working of the equipment must be given considerations in the design and selection of the equipments.
- All arrester used in this area must satisfy all or most of the requirements.
  - (i) Become conducting at voltage above diverter rating
  - (ii) Restrict the voltage across its terminals to the design value
  - (iii) Become none conducting again when the line to neutral voltage becomes lower than the design value, that is it should not permit any power follow-on current.
  - (iv) Not conducting any current at normal or some how above normal power frequency voltage.

- Routine but not frequent checks may be required to meet the "lightning index" used for designing the substations.

## 7.0 CONCLUSION

The Niger Delta region especially in the coastal areas the humidity is very high, almost throughout the year and rainfall period is about 9 months in the year with high keraunic level, therefore protection of substation requires careful selection of protective devices.

From the surge absorber known, it is necessary that the gapless ($Z_nO$) is best suited for the environment. From the foregoing it is proper to ascertain the protective ratio, the discharge current and the protective level of the arrester with the insulation level (BIL) of the equipment for proper co-ordination for reliability of the system.

Proper insulation co-ordination should ensure that the volt time curve of the weakest piece of equipment in the system will lie above the volt time curve of the protective device such as lightning arrester. The weakest piece is always the transformer, the single most expensive equipment with intricately formed insulation. Therefore the incoming voltage surge which would have otherwise damaged the transformer insulation is reduced to that of the volt-time curve (residual voltage) of the lightning arrester such that the transformer is fully protected.

## 8.0 REFERENCES


1. Begamudre, R.D., 2009: Extra High Voltage A.C Transmission Engrg: New Age International Publishers – New Delhi: 246 – 258.
2. CIGRE, 1991: Guide to procedures for estimating the lightning Performance of transmission lines W.C 01 (lightning) of SC 33 (over voltage and insulation co-ordination).
3. Dahiya, R.S., Attri, V., 2009: Substation Engineering Design Concept and Computer Application: S.K. Katarla and sons, Publishers of Engrg and Computer books; 467 – 487.
4. Ekonomou, L., Gonos, L.F., Stathopulus, I.A., 2003: An Improved Back flashover model for estimating the lightning performance of transmission level, UPEC, Thessoloniki, Greece: 107 – 112.
5. Gupta, J.B. 2008: A Course in Power Systems: S.K. Kataria $\propto$ sons, Publishers of Engrg and Computer books; Delhi.
6. Gustavo Carrasco, H., Alessandro Villa, R., 2003: Lightning Performance of Transmission Line, Las Claritas – Santa Elena up 230kv: International Conference on Power System Transient IPST 2003 in New Orleans U.S.A.
7. Martinez, J.A., Castro-Aranda, F. 2003: Lightning Performance analysis of transmission lines using the EMTP: Power Eng $S_{oc}$ Gen. meet. 1, 295 – 300.
8. Mousa, A.M. 1991: The Applicability of Lightning Elimination Devices to substation and Power Lines: IEEE Trans. On Power Delivery 13 (4), 1120 – 1127.
9. Mowete, A.I., Adelabu, M.A.K. 2009: Assessment of Lightning and other Electrostatic Disturbance along the coast of South West Nigeria 19th International Conference of Electromagnetic disturbance.
10. Nagrath, I.J., Kothari, D.P. 2001: Power System Engineering: Tata Mc Graw Hill Publishing Company Ltd New Delhi, 586 – 602.
11. Rao, S. 2008: Electrical Substation Engineering and Practice: Khanna Publishers Delhi, 187 – 210.
12. Razevig, D.V. 2003: High Voltage Engineering: Khanna Publishers, Delhi, 521 – 559.
13. United States Department of Agriculture, 2001: Design Guide for Rural Substations: RUS Bulletin 1724E – 300, http:/www.usda.gov/rus/electric
14. Uppa I, S.L., Rao, S. 2009: Electrical Power Systems: Khanna Publisher New Delhi: 1023 – 1054.
15. Wikipedia 2009: Electrical Substation.